Ankit Kumar Agarwal
Nikhil Sharma
Yogesh Kumar Sharma

Solar Panel Performance: The Effects of Water and Surfactant Cleaning

Ankit Kumar Agarwal
Nikhil Sharma
Yogesh Kumar Sharma

Solar Panel Performance: The Effects of Water and Surfactant Cleaning

LAP LAMBERT Academic Publishing

Imprint

Cover image: www.ingimage.com

Publisher:
LAP LAMBERT Academic Publishing
is a trademark of
Dodo Books Indian Ocean Ltd. and OmniScriptum S.R.L publishing group

120 High Road, East Finchley, London, N2 9ED, United Kingdom
Str. Armeneasca 28/1, office 1, Chisinau MD-2012, Republic of Moldova, Europe
Managing Directors: Ieva Konstantinova, Victoria Ursu
info@omniscriptum.com

Printed at: see last page
ISBN: 978-620-5-63161-4

TABLE OF CONTENT

LIST OF FIGURES

LIST OF TABLES

ABSTRACT

Around the world, solar photovoltaic (PV) panels are extensively used to convert solar radiation into electricity. But solar PV panels have adverse effect on conversion efficiency with the increase in dust deposition on PV panels. It was found that the deposition of only $4 g/m^2$ of dust reduce the power output by 40% [1]. Therefore cleaning of solar PV panels is necessary to achieve high productivity. There are lots of cleaning methods available. On the other hand, cleaning with surfactant is very effective and cost efficient. Many experimental and simulation studies are presented for solar PV panel cleaning. Different cleaning methodology has been analyzed in previous studies. It is found that use of surfactants has significant improvement on the conversion efficiency of PV panels.

In the present study the surfactant based cleaning and water based cleaning systems have been fabricated and both systems were compared with system without washing. The experimental investigations are done in the month of March, April and May (from 4th March to 1st June) to analyze the performance enhancement. The surfactant and water based system contains manual cleaning by mixing surfactant in water.

It is found that the efficiency of system without washing is decreased with a slope of 0.07/day. The efficiency of water based cleaning is decreased with a slope of 0.03/day while the efficiency of system washed with surfactant is consistent. The overall efficiency of System A was decreased by 29% and the efficiency of system B and system C was decreased by 18.17% and 5.2% respectively. Surfactant washing gives the consistent efficiency.

System A (without washing) generated 10.81 kWh power during the experiment. System B (washed with water) generated 11.95 kWh power while system C (washed with surfactant) generated 13.78 kWh power. System B generated 1.14 kWh more power as compared to system A. System C generated 2.97 kWh more power as compared to system A. System C generated maximum power output in expense of INR 1.274 for surfactants.

Chapter 1

INTRODUCTION

In India renewable energy sector is the fourth most attractive renewable energy market in the world as per the Renewable Energy Attractiveness Index 2018. Power generation from renewable energy sources in India reached 101.84 billion units in FY18 and 107.22 billion units during April 2018-January 2019. As of February 2019, total renewable power installed capacity (excluding large hydro) in the country stood approximately 77 GW. Overall, India is expected to add up to 8.5 GW of renewable energy capacity in 2018-19. The Government of India allocated Rs 3,004.90 crore (US$ 416.48 million) in the interim budget 2019-20 for development of solar power projects including both grid-interactive and off-grid and decentralized categories.

As India looks to meet its energy demand on its own, which is expected to reach 15,820 TWh by 2040, renewable energy is set to play an important role. By 2030, renewable sources are expected to help meet 40 % of India's power needs [2].

1.1 Solar energy: An important form of renewable energy

Energy obtained from sun is named as solar energy. Solar power in India is a fast growing industry. The country's solar installed capacity reached 253.94 MW on roof top and 1114.75 MW on ground mounted as of 30 June 2019.India has become globally the lowest cost producer of solar power [3]. Solar energy can be harnessed by two methods:

Photovoltaic solar technology

In this method the system directly converts sunlight into electricity using Photovoltaic (PV) panels made of semiconductor cells.

Solar thermal technology

In this method the system captures the heat of solar radiations. Then this heat is used directly (low-temperature solar thermal) or converted into mechanical energy and in turn electricity (concentrated solar power – CSP).

1.2 Photovoltaic solar technology

Photovoltaic solar technology includes solar energy conversion by means of solar cells. Solar cells are made of semiconductors used to convert the solar radiation (or other electromagnetic radiation) emitting on it into electricity. Fig. 1.1 shows the operating principle of a PV solar cell. When sunlight of a suitable wavelength falls on a solar cell, it separates into positive-negative charge carriers in enthralling material. In the existence of an electric field, it will produce a current in an external circuit, which gives voltage difference on terminal.

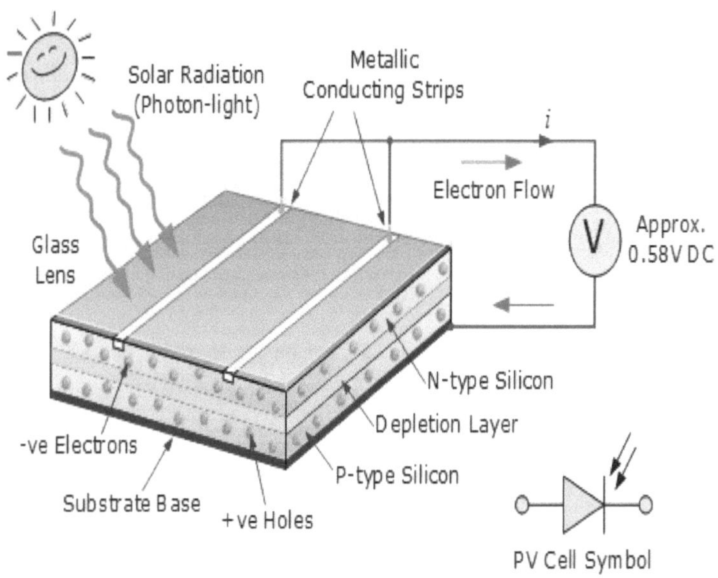

Figure 1.1 PV Panel working

1.3 Factors affecting energy conversion efficiency of PV panels

All of the sunlight that reaches on PV cell is not converted into electricity. Hence, the efficiency of PV panels is very low. To enhance the efficiency of PV panels effect of various factors that reduces the exposure of sunlight to PV panels are considered. The various factors are described as follows:

1.3.1 Effect of orientation

Wrong orientation could affect efficiency of solar power. If solar panels were not positioned in right direction rays hitting the panels will decrease that will definitely decreases its efficiency. To get maximum efficiency solar panels should be positioned to get maximum sunlight for maximum time. For getting maximum efficiency solar panels were mounted at latitude and longitude angle so that sunrays will hit the panel at 90°.

1.3.2 Effect of dust

The accumulation of dust on the solar cells panels decreases the efficiency of the solar cells day by day especially in the regions where rate of dust is high and frequency of rain is low. The accumulated dust on the solar cells panel blocks the sun's rays which decrease the performance of the solar cells over time until the cell panels are cleaned manually or by rain. So it is very important to clean the solar panels regularly to maintain good efficiency. There are lots of methods to clean solar panel.

1.3.3 Effect of temperature

A very high temperature can also lead to decrease in the output of solar panel thus can decreases its efficiency As the temperature rises current in solar cell increases but voltage of solar cell decreases. The decrease in voltage is more than that of increase in current so overall power output is reduced which results in decrease in efficiency.

1.3.4 Effect of type of solar cell

Efficiency also depends upon crystal structure of solar cell. Monocrystalline solar panels have solar cells made from a single crystal of silicon which makes them highly efficient while polycrystalline solar panels have solar cells made from many fragmented silicon melted together making their efficiency lesser than monocrystalline. Table 1.1 shows the different solar cells used for making solar panel with their efficiency and advantages and disadvantages [4].

Table 1.1: Solar Panel with their efficiency, advantages and disadvantages

Solar Cell Type	Efficiency-Rate	Advantages	Disadvantages
Monocrystalline Solar Panels (Mono-SI)	~20%	High efficiency rate; optimized for commercial use; high life-time value	Expensive
Polycrystalline Solar Pancls (p-Si)	~15%	Lower price	Sensitive to high temperatures; lower lifespan & slightly less space efficiency
Thin-Film: Amorphous Silicon Solar Panels (A-SI)	~7-10%	Relatively low costs; easy to produce & flexible	shorter warranties & lifespan
Concentrated PV Cell (CVP)	~41%	Very high performance & efficiency rate	Solar tracker & cooling system needed (to reach high efficiency rate)

The cleaning of solar panels plays important role in maintaining a significant efficiency of solar panels

1.4 Methods of cleaning of solar panel

Solar panels can be cleaned by various methods which were described below.

1.4.1 Cleaning by forced air

In this method solar panel is cleaned by a forced airflow. The forced air removes the dust particles from the surface of PV panels but this also requires more energy to produce pressurized air. Fig. 1.2 shows a conceptual diagram of this type of system.

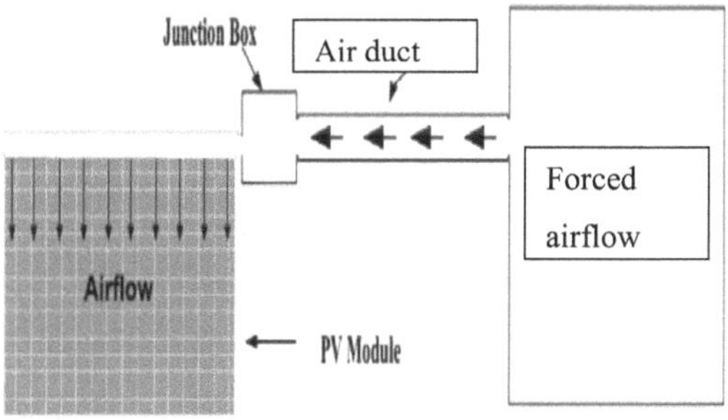

Figure 1.2: Conceptual diagram of forced air flow based cleaning

1.4.2 Water based cleaning

This method requires high pressure water in large amounts to remove soiling particulate matter stuck on the PV panel surface. The pressurized water requires high energy to generate high pressure. There are many drawbacks of this method, such as lack of reliability when applied in water scarce regions, loss of water in huge quantities and the attraction of more dust particles when the PV panel surface is wet after cleaning.

1.4.3 Mechanical cleaning

In this method set of mechanical instruments like robot or motors are required to operate the brushes or wipers and a water storage tank with sprinklers is used to clean the PV panel surface. This method uses automation where the system is controlled by a micro-controller with sensors. The system reduces labor cost for cleaning of solar panel surfaces. Its efficiency is uncertain and may prove to be not quite efficient in case of high dust accumulation. These drawbacks can be improved by adapting few measures to regulate the timing of cleaning to avoid thermal shocks, by using power efficient motors to operate brushes or wipers, and by using softer cleaning materials to avoid panel surface abrasion.

1.4.4 Cleaning by electrodynamic screens (EDS)

Cleaning by electrodynamic screen (EDS) placed on a solar panel can ensure automated and continuous removal of deposited dry dust without the use of water or any mechanical moving parts. It is an active dust removal method that utilizes high voltage supplies to power electrodes to generate an electric field which helps in removing the charged and uncharged dust particulate matter from the PV panel by shifting them over an edge of the panel surface. EDS screen cleaning is shown in Fig 1.3.

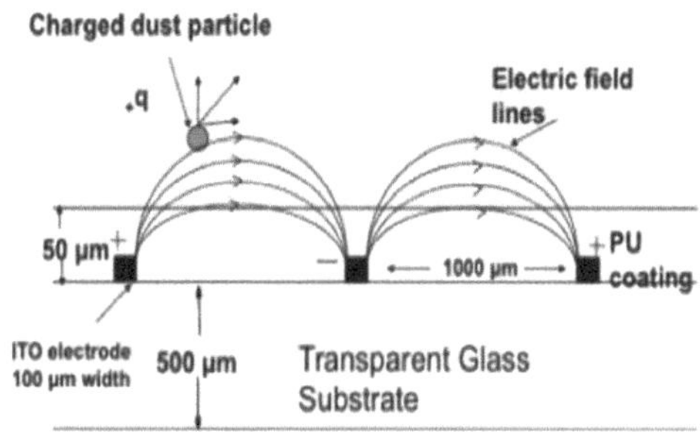

Figure 1.3 EDS screen cleaning

1.4.5 Self-cleaning coatings

In this method a chemical coating and nano-patterned fabrication is done on glass surface. This method utilizes tailor-made hydrophobic or hydrophilic surface by chemical coating/screen layer. The coating doesn't allow dust to stick on the surface of panels and chemically breakdown the organic dirt through the reaction to UV rays. Degradation of coating is the main drawback of self-cleaning coatings.

1.4.6 Cleaning using surfactants

In this method the surface is cleaned by using water mixed with a surfactant. Surfactant is a wetting agent that lowers the surface tension of water and allows water easier spreading. This is an improvised version of water based cleaning system. As the dust particles contain positive, negative and neutral particles surfactants are capable of removing these charged and uncharged particles as well.

1.4.7 Rainfall cleaning

This is a natural process of cleaning. This process usually leaves dust that is stuck onto the panels due to moisture, and requires heavier rainfall to remove it. For getting consistent efficiency in power generation the requirement is complete and timely cleaning of solar panels.

1.5 General Description of Research Concept

Around the world, solar photovoltaic (PV) panels are extensively used to convert solar radiation into voltage output, further into electricity. But solar PV panels have an adverse effect on conversion efficiency with the dust deposition. This drop may up to5–6% decrease in the PV efficiency in 1 week depending on dust concentration on solar cell [1]. Therefore cleaning of solar PV panels is necessary to achieve high productivity. There are lot of cleaning methods available which were discussed above. Cleaning with surfactant involves washing of solar panel by mixing surfactant with water.

1.6 Objectives of the Thesis

The major objective of this present work is to test the efficiency of solar PV panel using different cleaning methods. Sub-objectives of this thesis are:

- Design and fabrication of experimental setup which includes three systems as reference solar panel cleaning by surfactant, panel cleaning with water, no cleaning.
- Experimentally investigating the performance of solar panels by calculating the power output and efficiency of all systems.
- Comparing and analyzing the power output and efficiency obtained of all three systems.

1.7 Thesis Outline

The organization of following chapters is as follows:

Chapter 2: This chapter describes literature review of the knowledge that is previously available including the problem of concern. The research work is presented on the performance improvement of solar PV panel using different cleaning techniques.

Chapter 3: This chapter deals with material selection, experimental setup design and method of experiment. It gives the description of surfactant which includes working principle, comparison and selection criteria. It also gives the complete description of components along with technical specifications and fabrication process. The chapter is ended with method of experiment in different steps.

Chapter 4: The chapter shows the result and discussion for voltage, current, power output gain, efficiency and economical analysis.

Chapter 5: Represents summary and conclusions from the experimental approach and future recommendation for further research.

Chapter 2

LITERATURE REVIEW

This chapter deals with the past research done and published in the field of cleaning of solar panels by various methods. The effect of cleaning of solar PV panels on the performance and surface of panels are also represented. Tabular reviews are also represented for better understanding.

2.1 Cleaning by mechanical instruments

Patil P.A. et al. [5] mathematically optimized the dust cleaning equipment. The optimization is done on cleaning brush design. The experiment concluded the optimized brush increased the absolute efficiency of panel from 9 to 11.6%.

Thomas S.K. et al. [6] experimentally installed a solar panel cleaning system that can be used with/without water. The system uses brushes made up of nylon to clean the panels that has two squeegee on both sides of brush to clean water from panel .system can be fixed directly on to the panels in solar power plant, commercial, residential sectors. The experiment concluded that the water can be reused and system increased the efficiency of solar panel by removing all kinds of dust particles also in future the system can be upgraded in fully automated system.

Parrott B. et al. [7] studied about dry cleaning of solar panel from brushes. The study included a new type of brush, which uses silicone rubber foam flaps mounted onto an aluminum core. The study concluded that silicone rubber is more effective in minimizing the impact of dust on the solar panels and does not induce any damage to the surface of the solar panels.

Table: 2.1 Literature Review summary for mechanical instruments

Reference	Authors	Method of cleaning	Observation
[5]	Patil P.A. et. al.	Used mechanical brushes	Optimized design of brush used in cleaning.

Reference	Authors	Method of cleaning	Observation
[6]	Thomas S.K. et. al.	apparatus that utilizes a nylon spiral brush and two wipers cleaning system were developed.	The final prototype was tested successfully and produced satisfactory results.
[7]	Parrott B. et. al.	Developed a silicone rubber foam brush on an automated robotic cleaning tool	The study indicates that robotic cleaning systems utilizing silicone rubber foam brushes show promise for waterless dust mitigation activities.

2.2 Cleaning by electrodynamic screen (EDS)

Kawamoto H. [8] experimentally studied the Electrostatic cleaning equipment for dust removal from soiled solar panels. By application of high AC voltage between the parallel screen electrodes of the device, the resultant Coulomb and dielectrophoresis forces act on the dust particles near the electrodes. The dust particles are agitated by the alternating electrostatic field near the electrodes. The experimental results showed that approximately 80% of the accumulated dust is cleaned when the inclination is higher than 20°. High performance was achieved by applying a high voltage; however, saturation occurred at a high value. The applied voltage was limited by the insulation breakdown. The power consumption of the system is negligibly low.

Deb D. and Brahmbhatt N.L [9]. studied the various methods of solar panel cleaning which includes Forced flow from air, Rainfall cleaning , Water-based cleaning , Manual cleaning , Mechanized cleaning , Electrodynamic screens (EDS),Super Hydrophilic plane (SHIP),Super Hydrophilic plane (SHIP) and then proposed a water free automated cleaning which uses a machine for cleaning panels .the machine uses brushes for cleaning panels. The study concluded that solar panels need cleaning, and depending on the environmental conditions of installations, appropriate cleaning mechanism can be applied.

Kawamoto H. and Guo B. [10] proposed improved cleaning system for dust on solar panel. The experiment used Namib sand and Doha dust. The high voltage is generated by positive and negative amplifiers switches and applied between the thin cover glass and glass substrate.

The set up concluded that high voltage improves the cleaning performance also insulation strength of cleaning plate must be improved to handle high voltage.

Mazumder M.K. et al. [11] reviewed the reviews the progress of the electrodynamic screen (EDS) film technology for frequent water-free cleaning with low energy requirements. EDS film consists of rows of interdigitated parallel electrodes embedded within two transparent dielectric films. The study concluded that EDS films can be a effective way for cleaning solar panel without water.

Sharma R. et al. [12] experimentally studied the performance of EDS screen for cleaning of solar panel for mars mission. The experiment used two types of EDS screen which are a transparent EDS by direct laser patterning of ITO electrodes on polyethylene terephthalate (PET) substrate and the printed-circuit-board EDS with copper electrodes was fabricated by photolithography. In both cases, the electrodes were embedded in a transparent polyurethane film. The study concluded the increment of dust removal efficiency (DRE) by 90% in high dust enviourment . The transmittance for ITO/PET screen was 83%, it decreased to 58% when ITO electrodes were etched using laser ablation. When placed on top of the solar panels, the transparent screens decreased the power output of solar panels by 15%.

Kawamoto H. [13] experimentally studied performance of EDS screen placed on the conveyer belt. The conveyer is made up of parallel copper electrodes etched by photolithography on a plastic substrate. Numerical simulation of experiment is performed and resulted that. Lunar soil was successfully transported by EDS conveyer created on the conveyer. But, very small particles less than 1 Pm and large particles were not transported.

Mazumder M.K. et al. [14] theoretically and experimentally studied the dust removal efficiency of EDS screen and found that EDS screen made up of transparent plastic such as PET for its UV radiation resistance, and a set of parallel conducting electrodes, made of transparent indium tin oxide (ITO) or a conducting polymer, embedded under a thin transparent film on the surface provides an effective shield against dust deposition and obscuration of the panels from solar radiation. The experiment also concluded that EDS film technology is efficient for protecting solar panels on surface of mars.

Table: 2.2 Literature Review summary for EDS cleaning

Reference	Authors	Method of cleaning	Observation
[8]	Kawamoto H.	High AC voltage is applied between the parallel screen electrodes of the device. The dust particles are agitated by the alternating electrostatic field near the electrodes	We observed that high performance was achieved by applying a high voltage; however, saturation occurred at a high value. The applied voltage was limited by the insulation breakdown.
[9]	Deb D. and Brahmbhatt N.L.	Review of soil prevention methods is presented with a view to increase the energy capture from solar panels.	Studies suggest that electrostatic cleaning solutions consume less power and are effective only for dry and tilted panel surfaces also solar panels need cleaning, and depending on the environmental conditions of installations, appropriate cleaning mechanism can be deployed
[10]	Kawamoto H. and Guo B.	An improved cleaning system has been developed that uses electrostatic forces to remove dust from the surface of solar panels	High voltage improves the cleaning performance also insulation strength of cleaning plate must be improved to handle high voltage

Reference	Authors	Method of cleaning	Observation
[11]	Mazumder M.K. et.al.	EDS screens were used for cleaning	Studies show that a transparent EDS film can be used to laminate solar PV modules and CSP mirrors for self-cleaning function without requiring water.
[12]	Sharma R. et.al.	The electrodynamic screen (EDS) is considered to be one of the feasible dust mitigation technologies for future Mars missions	EDS was successful in dust removal even with these high dust loadings corresponding to more than a one-year worth of dust deposition on the surface of Mars.
[13]	Kawamoto H.	Non-uniform electrostatic field was applied to the particle near the tip of the electrode.	It was demonstrated that single particle of the lunar soil was manipulated with an electrostatic dipole probe. However, small particles could not be released even when the voltage application was turned-off due to the adhesion force and/or the Coulomb force.

Reference	Authors	Method of cleaning	Observation
[14]	Mazumder M.K. et. al.	A dust cloud chamber with appropriate ventilation was used for testing the EDS.	Three-phase high voltage AC electric field to the electrodes, electrodynamic screens repel and remove dust particles from the screens regardless of whether the dust particles are initially charged or uncharged.

2.3 Cleaning by self cleaning coatings

Latthe S.S.et al. [15] studied the self cleaning coating applications by applying the coating on various surfaces. First all the substrates were cleaned by water and detergent to remove dust and organic components then hydrophobic silica nanoparticles were simply dispersed in hexane (5 mg/ml), and ultrasonicated for 30 min. and applied on various substrates by dip or spray coating technique depending on the shape of the substrate. While the dip coating process, the effect of deposition time and deposition layers on the wettability of the coatings was studied in detail. The 10 layers were applied with the deposition time of 1 min for each deposition layer. Whereas during spray coating, 12 spray layers from the distance of 15 cm were sufficient to achieve superhydrophobicity. After deposition, the coatings were dried overnight at room temperature. The experiment concluded that the coating on every substrate showed the self cleaning ability.

Mozumder M.S. et al. [16] reviewed the self cleaning coatings for solar panel cleaning. Superhydrohphobic and superhydrophillic coatings were studied in detail. Methods like Solvent based methods (Sol-gel method, Dip coating, Electro spinning) Vapor-assisted coating methods(Plasma etching, Chemical vapor deposition), Dry powder coatings were studied to apply coatings on solar panel.

Zhan W. et al. [17] experimentally studied the self cleaning coating for cleaning solar panels. The experiment consist of Borosilicate glass comprising of Na2O (8 wt%), B2O3 (23 wt%), and SiO_2 (69 wt%) was annealed in air at temperatures ranging from 600 to 720 °C for 30 min to 24 h to allow phase separation into an interpenetrating pattern consisting of alkaliborate and silica-rich phases. After heat treatment, the surface composition of the glass becomes enriched in SiO_2 due to volatilization of sodium oxide and boron oxide, which has a significant effect on the subsequent acid etching process. The phase-separated glass was immersed into a dilute mixture of deionized water and 6:1 homemade buffered oxide etchant (v/v, 50:1) at 25 °C for 2–5 min for acid pretreatment. Then the porous structure was formed under acid etching with ultrasonic activation in a dilute mixture of 100 ml of deionized water and 5 ml of 98% sulfuric acid solution and 5 ml of 40% nitric acid for 5–10 min. Finally, the glass was rinsed with deionized water to remove the residual acid solution on the surface. The experiment concluded that by varying phase separation and acid treatment processes for borosilicate glass, the surface nanostructure can be controlled to achieve substantial gains in light transmission.

Adak D. et al. [18] experimentally developed a highly transparent, self-ordered, superhydrophilic and photoactive TiO_2 thin film coatings by sol-gel based self-assembly. To improve the performance of TiO_2 thin film, film was treated under nitrogen plasma known as plasma based approach. The experiment concluded that Coatings are superhydrophilic also Coatings are highly transparent with optical transmission as high as 95% with water contact angle $< 5°$. Finally, coatings are found to be mechanically robust with micro hardness in the range of 0.95–2.02 GPa.

Walz K.A. et al. [19] experimentally studied the effect of self cleaning coating which comprises of titanium oxide (TiO_2) and silicon oxide (SiO_2) with solution gel chemical processing. The study demonstrated a successful field application process for the deposition of self-cleaning coatings for solar panels. The experiment didn't concluded any result but showed that self cleaning coating can be studied more to improve the efficiency of solar panel.

Tao C. et al. [20] experimentally deposited the double layer TiO_2 SiO_2 coating calcined at 550 C for three hours using the SiO_2 sol at a withdraw rate of 80 mm per minute. Then coating was treated by SSP for one cycle. A porous open pore-structure mono-layer SiO_2 coating was fabricated. The experiment resulted that the coating shows transmittance, hydrophilicity, photocatalytic activity, mechanical robustness and long-term stability against high temperature and moisture. The TiO_2 and SiO_2 coating can be further studied to improve solar panel efficiency.

Arabatzis I. et al. [21] deposited the liquid SSG coating with the thickness of 150mm. The SSG coating is water based suspension which after application on glass creates an inorganic and nano-structured coating chemically bonded to the substrate. Surface Shield G (SSG) contains metal oxides mixture mainly of titanium oxide (TiO_2) anchored with silica (SiO_2). The experiment concluded that the coatings showed excellent stability in thermal and UV weathering and coatings were not drastic in sandblasting conditions. Coated glass were better in comparison to the uncoated glass showed better antireflective and self-cleaning toward dust properties.

S. Sutha et al. [22] experimentally prepared a superhydrophobic coating by mixing aluminum nitrate nonahydrate with 2-methoxyethanol solution by magnetic stirring in the presence of monoethanolamine (MEA) stabilizer at room temperature for 30 min. the experiment resulted that coating is optically transparent , anti-reflective and superhydrophobic.

Mehmood U. et al. [23] reviewed the basic concepts and principles of antireflection and self-cleaning coatings their detailed fabrication. Review also contains challenges faced in practical applications of self cleaning coatings and their future developments. Overall, an overview of antireflective surfaces with super hydrophobic characteristics is provided in light of the current challenges.

Daniel Ebert and Bharat Bhushan [24] experimentally created rough surfaces on PDMS substrates using a O_2/CF_4 plasma by deep reactive ion etching (DRIE) .after that an additional treatment with either octafluorocyclobutane (C_4F_8) plasma or vapor deposition of perfluorooctyltrichlorosilane (PFOTCS) following surface activation with O_2 plasma was done. The effects of surface roughness and the additional surface modifications were examined with respect to the contact angle, contact angle hysteresis, and optical transmittance. To examine wear resistance, a sliding wear experiment was performed using an atomic force microscope (AFM).

Powell M.J. et al. [25] experimentally deposited a multifunctional $VO_2/SiO_2/$ TiO_2 film on glass substrate. The experiment uses atmospheric-pressure chemical vapor deposition method, vibrational spectroscopy and electron microscopy techniques. The deposited film exhibited excellent solar modulation properties as well as high transparency and resistance to abrasion, compared to single VO2 films of the same thickness. Self-cleaning properties were also observed. The SiO2 acted as a barrier layer, preventing the diffusion of Ti4+ ions into the VO2 layer. The SiO2 also promoted the optical properties and allowed for superior thermo chromic behavior when compared to single VO2 films. The experiment resulted deposition of the SiO2/TiO2 over layer showed improvement of visible-light transmission.

Xu Q.F. et al. [26] studied the wetting properties, optical properties ,durability and anti soiling behavior of a single high performance polymeric coating that shows both anti reflective and anti soiling properties. The studied concluded that coating will outperform in dusty and high soiling environment.

Ragesh P. et al. [27] reviewed the importance of self-cleaning coating materials. Coatings such as oleophobic coatings, amphiphobic coatings and multifunctional coatings, have also been reviewed. The study also reviewed the methods of fabricating super hydro phobic and super hydrophilic coatings.

He G. et al. [28] reviewed the cleaning method for solar panel array which includes Natural removal of dusts method, Mechanical removal of dusts, Self-cleaning nano-film, and Electrostatic removal of dusts. Study concluded that best method of removal-dusts for solar cell array is electric curtain.

Verma L.K. et al. [29] experimentally studied Nanostructuring process of solar cell packaging which allows an enhancement in solar cell performance due to an increase in net optical transmission through the nanostructured glass. The experiment concluded that glass surface becomes superhydrophiic after nanostructuring.

Table 2.3: Literature Review summary for self cleaning coatings

Reference	Authors	Method of cleaning	Observation
[15]	Latthe S.S et.al.	Hydrophobic silica nanoparticles were simply dispersed in hexane (5 mg/ml), ultrasonicated for 30 min. and applied on	The dip and spray deposition layers were optimized to achieve the superhydrophobicity.

Reference	Authors	Method of cleaning	Observation
		various substrates by dip or spray coating technique.	
[16]	Mozumder M.S. et.al.	Use of TiO_2 film as superhydrohobic plane.	Self-cleaning coatings ease the removal of dust from the solar panels that in turn increases their energy conversion efficiency.
[17]	Zhan W. et. al.	Fabrication of nanostructured coating	Smaller particles that are not removed by the action of vibration, gravity and wind, can be cleared via self-jumping of water droplets
[18]	Adak et. al.	The morphologies and surface roughness of TiO_2 coatings were examined	The dirt rides piggyback and washes away with water as water easily seeps in between the superhydrophilic coating and attached dirt particles.
[19]	Walz K.A et. al.	coating materials used in this experiment were deposited from a nanoparticulate suspension containing a mixture of titania, TiO_2, and silica, SiO_2 metal oxides	This study demonstrated a successful field application process for the deposition of self-cleaning coatings for solar photovoltaic panels
[20]	Tao C. et. al.	Silica (SiO_2) and titanium dioxide (TiO_2) nanomaterials were used as building blocks to construct multifunctional AR coatings	developed a novel, convenient technique to prepare multi-functional coatings by the utilization of TiO_2eSiO_2

Reference	Authors	Method of cleaning	Observation
[21]	Arabatzis I. et.al.	The liquid SSG product was deposited on the glass substrates by sprayer	The θ value was reduced from 39° on the uncoated glass to 6° on the coated surface demonstrating super hydrophilic property of the latter.
[22]	Sutha S. et. al.	super hydrophobic surfaces based on aluminum oxide coatings had been developed on glass substrates via solution based approach	The 300 nm thick nanoflakes like interconnected network of coatings with 38% of porosity enabled the coating to acquire the competitive properties of superhydrophobicity with static water contact angle exceeded 160° and average transmittance level of 95%
[23]	Mehmood U. et. al.	Self-cleaning surfaces exhibit special anti-wetting properties owing to the water contact angle greater than 150° leading to the ready rolling-off of water droplets	Further investigations of surface hydrophobicity are required to include atmospheric effects such as temperature, humidity, icing, etc., and dust accumulations and adhesion on hydrophobic surfaces in calculations and modelling.
[24]	Ebert D. and Bhushan B.	deep reactive ion etching (DRIE) was used to create rough surfaces on PDMS substrates using a O2/CF4 plasma	superhydrophobic surfaces with high transparency were created on PDMS with a simple process using deep reactive ion etching and subsequent surface modification
[25]	Powell M.J. et. al.	multifunctional, robust, layered VO2/SiO2/TiO2	the multilayer VO2/SiO2/TiO2 films showed strong adherence

Reference	Authors	Method of cleaning	Observation
		films were quickly deposited on glass substrates using atmospheric-pressure chemical vapor deposition	to the glass substrates
[26]	Xu Q.F et. al.	Studied an anti-reflective and anti-soiling coating for photovoltaic panels	Coating will outperform in dusty and high soiling environment.
[27]	Ragesh P. et. al.	self cleaning materials were discussed	Smart self-cleaning coatings are those which respond to external infuences such as electric feld, temperature, and light
[28]	He G.et. al.	The methods of dust-removal were discussed.	the best strategy of removal-dusts for solar cell array is electric curtain
[29]	Verma L.K. et.al.	process of non-lithographic nanostructuring of the packaging glass surface is shown	Nanostructuring allows an enhancement in solar cell performance due to an increase in net optical transmission through the nanostructured glass.

2.4 Cleaning by other methods

Williams R.B. et al. [30] experimentally studied the solar panel cleaning system by using piezoelectric system i.e. solar panel equipped with piezoceramic actuation were tested. Study concluded that result shows that mechanical vibration can remove dust.

Makkar A.et al. [31] studied the effects of dust on solar panel performance. The study concluded that efficiency of solar panel reduced with increase in dust deposition. Cleaning can significantly increases the solar panel efficiency.

Yilbas B.S. et al. [32] studied the materials used in making super hydrophobic layers. Some basic definitions and background for surface hydrophobicity are introduced. Later, case studies are incorporated to account for the processes involved with surface hydrophobicity.

K. A. Moharram et al. [1] experimentally studied the effect of using surfactant in cleaning of solar PV panels. The study compared the results with cleaning of PV panels with water and concluded that surfactant cleaning increases the efficiency of solar panels significantly.

Gaio J.N. [33] studied the ideal time for cleaning photovoltaic panels, aiming at high system productivity. When analyzing the data obtained in the GO, the following averages were found 70.83% on Performance Ratio; 12.96% on Capacity Factor; and 1111.24 kWh on annual yield. In order to determine the optimal time for the achievement cleaning time of the GO photovoltaic panels, it was established as a standard that the system would operate regularly with a performance ratio of at least 70%. If all the months when there was a performance ratio lower than 70% were changed so as to have at least 70%, the new average would be 72.46%. In this way, it was verified that the rate's decay over the years, in the first moment was seven months, in the second ten months, and eleven months in the third situation. Thus, an average which defined that between nine and ten months is the ideal time to clean the GO´s photovoltaic panel has been defined.

Nižetić S.et al. [34] experimentally studied an alternative cooling technique for photovoltaic (PV) panels that includes a water spray application over panel surfaces. An alternative cooling technique in the sense that both sides of the PV panel were cooled simultaneously, to investigate the total water spray cooling effect on the PV panel performance in circumstances of peak solar irradiation levels. The study concluded that average panel temperature was reduced from 52°C up to an average 24°C, and the lowest possible panel temperature was limited due to constant pipeline water temperature that was around 17°C.

Table: 2.4 Literature Review summary for other methods:

Reference	Authors	Method of cleaning	Observation
[30]	Williams R.B. et. al.	Solar panel equppied with piezoceramic actuation were tested	result shows that mechanical vibration can remove dust
[31]	Makkar A et. al.	Objective is not cleaning but it analyses that dust reduces the performance of solar panel.	A methodology is required to intimate the proper cleaning time of solar panels.
[32]	Yilbas B.S et. al.	Basic definitions and background for surface hydrophobicity are introduced.	Super hydrohobicity increases solar panel efficiency by decreasing the contact angel between surface and dust.
[1]	K. A. Moharram et al.	Surfactant cleaning	Surfactant cleaning system maintains the efficiency.
[33]	Gaio J.N et. al.	Objective of this work is to determine the ideal time for cleaning photovoltaic panels.	An average time is between nine and ten months is the ideal time to clean the GO´s photovoltaic panel has been defined.
[34]	Nižetić S et. al.	Alternative cooling technique for photovoltaic (PV) panels that includes water spray application over panel surfaces.	The best cooling option turned out to be simultaneous cooling of front and backside PV panel surfaces.

2.5 Conclusions of literature survey

From the literature review survey it can be concluded that

- Mechanical cleaning requires extra instruments for cleaning. It also requires extra power for cleaning PV panels.
- EDS cleaning system also requires high power to clean PV panels also at high voltage may cause insulation breakdown of panels.
- Self cleaning coatings faces the coating deteoration issues.
- The cleaning of PV panels by mixing of surfactant with water is easier method and enhances the performance of panels significantly. The requirement of water is also less than other methods.

Thus this study presents the motivation for comparing results by surfactant cleaning as there is no extra power required for cleaning and PV panel surface is also not affected by surfactant cleaning.

Chapter 3

MATERIAL AND METHODS

In this chapter the description of various materials and methods used for performing experiment are presented.

3.1 Experimental setup

The experimental setup mainly consists of 3 PV panels, panel cleaning unit and measuring instruments. Panel cleaning unit consists of 2 water tank, PVC pipe, connection and surfactant. The schematic diagram and pictorial view of experimental setup is shown in Fig. 3.1 and Fig. 3.2 respectively.

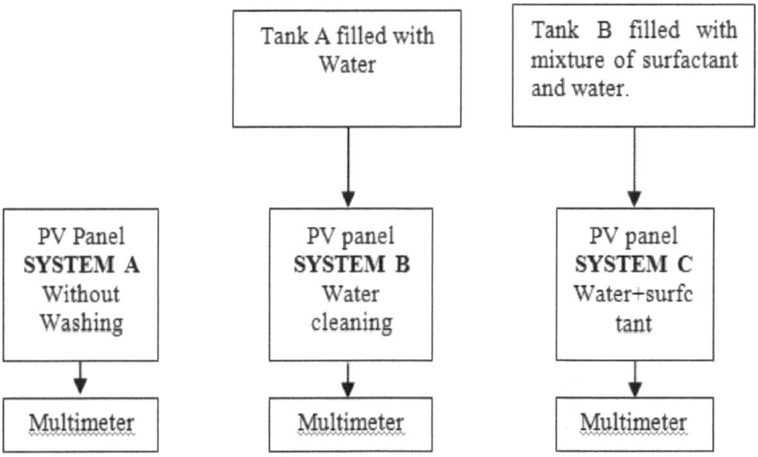

Figure 3.1 Schematic diagram of experiment setup

Figure 3. 2 Pictorial view of experimental setup

3.1.1 Solar photovoltaic panel

A Polycrystalline PV panel (36 Polycrystalline solar cells, model ELDORA 20P) of dimensions 552 mm × 354 mm × 18 mm with power rating 20W is used for research work. This PV panel is manufacture by Vikram Solar Ltd. West Bengal, India. The front and back view of panel is shown in Fig.3.3

Figure 3.3 Front and back view of PV panel (ELDORA 20P)

The solar panels were mounted at 26° inclined (equals to latitude of Jaipur which is 26.9°) from horizontal plane faces towards south to get maximum solar.

Figure 3.4 Mounting of solar panels at 26° angle

The technical data sheet including electrical, temperature coefficients and mechanical characteristics is shown in table 3.1

Table 3.1 Technical Data sheet of PV panel (ALDORA 20P),

All Data refers to STC (AM1.5, 1000W/m^2, 25°C)

Characteristics	Value/Tested
Peak Power (0-4.99Wp) Pmax (Wp)	20
Open Circuit Voltage V_{oc} (V)	21.44/22.18
Short Circuit Current I_{sc} (A)	1.27/1.46
Maximum Voltage V_{mpp} (V)	17.15/18.03
Maximum Current I_{mpp} (A)	1.18/1.36
Module Efficiency (%)	10.23/12.54
Temperature coefficients (T_c) of Open Circuit Voltage (β)	-0.31%/°C
Temperature coefficients (T_c) of Short Circuit	0.058%/°C

Characteristics	Value/Tested
Current (α)	
Fill Factor	0.74/0.76
Length × Width × Height	552 mm × 354 mm × 18 mm
Weight	2.8Kg
Cells	36 Polycrystalline solar cells, 2BB/3BB
Frame	Anodized aluminium frame with twin wall profile
Application Class	Class A (Safety class II)

3.1.2 Surfactants

Surfactants are wetting agents that lower the surface tension of a liquid, allowing easier spreading, and can also lower the interfacial tension between two liquids. . The term surfactant was coined by Antara Products in 1950 [35]. The word surfactant is short for "Surface Active Agent." surfactants are chemicals that, when dissolved in water or another solvent, orient themselves at the interface (boundary) between the liquid and a solid (dirt), and modify the properties of the interface (solid and liquid).

Surfactant molecule consists of two ends i.e. head and a tail .One end of the molecule has a long nonpolar chain that is attracted to oil, grease, and dirt (the hydrophobic). Another part of the molecule is attracted to water (the hydrophile). The surfactant lines up at the interface as shown in Fig. 3.5. The hydrophobic end of the molecule gets away from the water and the hydrophilic end stays next to the water. When dirt or grease is present (hydrophobic in nature) the surfactants surround it until it is dislodged from the boundary. Working behavior of surfactant is shown in Fig. 3.5.

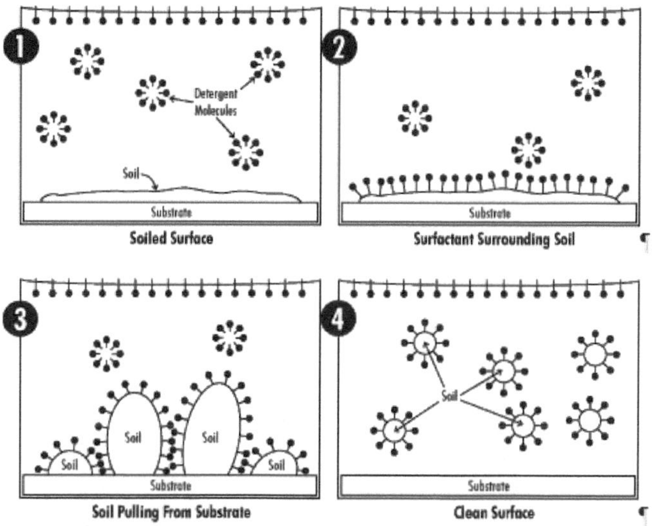

Figure 3.5 Working principal of surfactant

Surfactants are broadly classified into three types

1. Anionic surfactants – they are negatively charged surfactants they repel negatively charged dust particles like sand.

2. Cationic surfactants – they are positively charged surfactants and repel positively charged dust particles like carbon.

3. Zwitterionic surfactant – they are mixture of both anionic and cationic surfactants. They repel dust particles irrespective of charge.

We select the zwitterionic surfactant for our experiment as the dust particles contain both positive and negative charged particles so zwitterionic surfactant will give the best result. The composition of zwitterionic surfactant is

Sodium dodecyl sulfate ($NaC_{12}H_{25}SO_4$) + Cetylpyridinium bromide ($C_{21}H_{38}BrN$)

↓

Zwitterionic surfactant

Calculating the CMC (critical micelle concentrations) values of both surfactants the concentration is chosen as 1 gm of surfactant in 2 liter of water for experiment. The surfactant is shown in Fig 3.6.

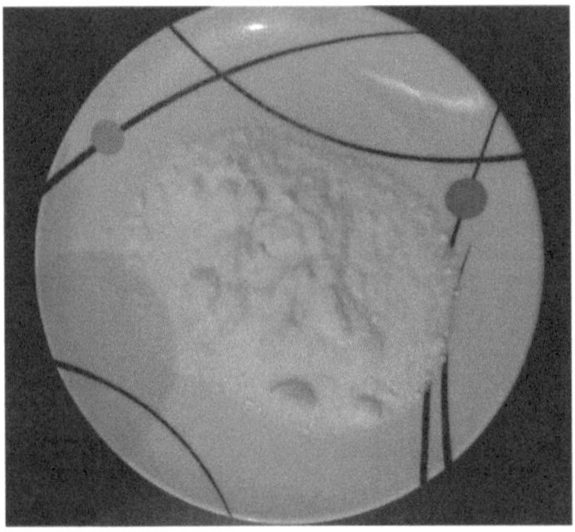

Figure 3.6 Surfactant

3.1.3 Multimeter

Multimeter is used to measure open circuit voltage and short circuit current of all three PV panels. The open circuit voltage is measured at 20 V range and the short circuit current is measured at 20 mA range. Fig. 3.7 represents the pictorial view of multimeter and table 3.2 represents technical data sheet of multimeter.

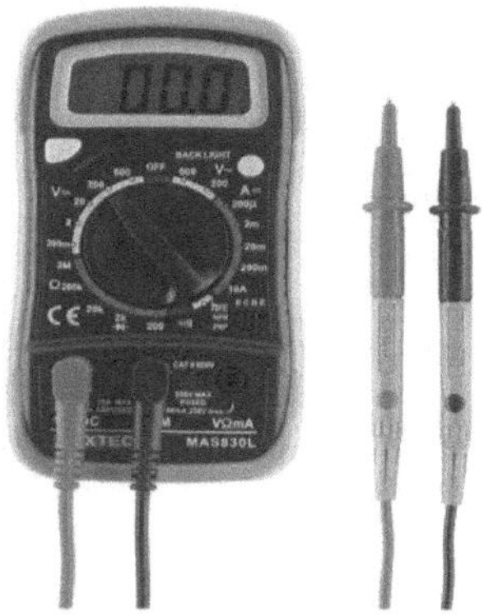

Figure 3.7 Pictorial view of Multimeter

Table 3.2 Technical data sheet of Multimeter

Parameters	Range
DC voltage range	200mV-1000V
DC voltage accuracy	+- 1.5%
AC voltage range	0-750 V
AC voltage accuracy	+-1.8%
DC current	2mA to 20A
DC current accuracy	+-2%
Resistance range	0-200 ohm
Resistance accuracy	+-1.5%

3.1.4 Panel cleaning unit

Panel cleaning unit consist of 2 separate water tanks one with water and another with mixture of water with surfactant and PVC pipe. The water used for single wash is 2L and the panels were washed with natural pressure.

3.2 Methodology

An outline of experiment procedure is shown in Fig.3.7.in the form of a flow chart, which describes the way in which present research work is to be undertaken. The method of experiment is described as follows:

Three panels were taken for performing experiment. Panel A is for without washing system, Panel B is for washing by water and panel C is for washing surfactant mixed in water.

These panels are mounted on the roof top at 26° angle to the horizontal surface facing towards south. Two water tanks and PVC connections are used for cleaning of panels B and C. One water tank is filled with water to clean panel B and another water tank is filled with mixture of water and surfactant with concentration of 1gm in 2 liter to clean panel C.

A multimeter is used to measure voltage (V_{oc}) and current (I_{sc}) after an interval of every 1hour from 9 am to 5 pm daily. Then voltage and current readings were averaged out to get one voltage and current data for one day.

The average power generation per hour is calculated by multiplying the daily average current, average voltage and fill factor (0.76 constant). The formula for daily power is given in equation 1.

$$\text{Average power generated per hour} = V_{oc} * I_{sc} * \text{Fill factor} \qquad \text{equation (1)}$$

The data of solar radiation intensity (G) is taken from MNIT whether data [36] on which the data are available at interval of every 15 minutes daily.

Total generated power is than calculated by multiplying the average power generated per hour by total time of experiment (8 hours) and then summing all the powers obtained.

The efficiency is the calculated by formula given in equation 2

$$\text{Efficiency} = \frac{\text{daily avg. power generated}}{\text{Avg. daily solar radiation * area of panel}} \quad \text{equation (2)}$$

The panels were cleaned at interval of every 7 days. Table 3.3 represents the dates on which cleaning is done manually and dates on which the rainfall occurred.

Table 3.3 Dates of cleaning panels

Washing count	Dates of cleaning
Start of experiment	4-3-2019
1	7-3-2019
2	14-3-2019
3	21-3-2019
4	28-3-2019
5	4-4-2019
6	11-4-2019
7	18-4-2019
8	25-4-2019
9	2-5-2019
Rainfall	3-5-2019
10	9-5-2019
11	16-5-2019
12	23-5-2019
13	30-5-2019
End of experiment	1-6-2019

The total power generated during the experiment i.e. from 4th March to 1st June 2019 for all three panels were calculated. The difference between the powers generated by panels were calculated and represented in results.

The economic analysis is also done for cleaning with surfactants.

Flow chart of experiment is shown in fig 3.8.

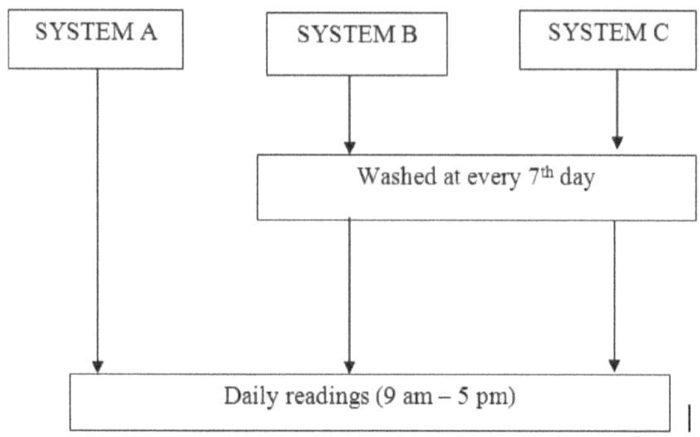

Figure 3.8 Outline of experiment procedure

Chapter 4

RESULTS AND DISCUSSION

This chapter presents the results obtained by the system during experiment dated from 4th March 2019 to 1st June 2019. Based on the results obtained, discussions of results are also presented.

4.1 Solar Radiation Intensity (G)

The radiation received from the sun to earth's surface is converted to electricity directly by the Photovoltaic Panels. The generation of electricity is dependent on the intensity of radiation received by panel. The generation of electricity increases when solar radiations increases and decreases when solar radiations decreases due to weather condition.

In this experiment solar radiation data is taken from MNIT weather data [36]. Fig. 4.1 shows variation of solar radiations falling on surface during the experiment days.

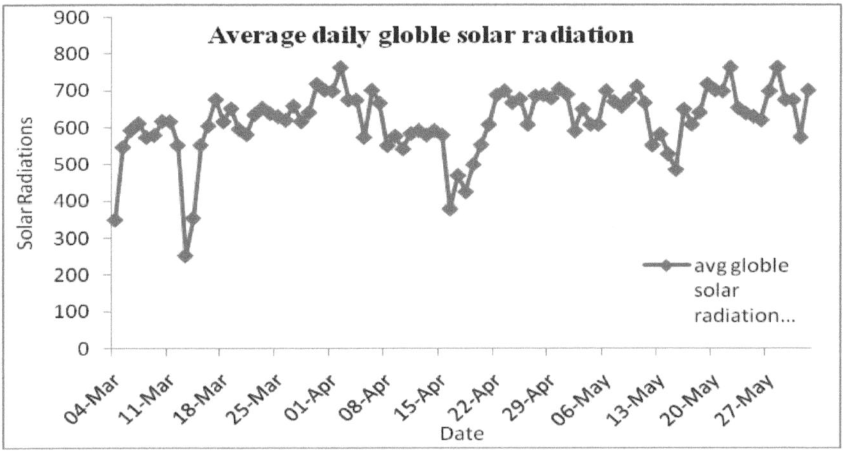

Figure 4.1 Daily average solar radiations

It is observed from the Fig 4.1 that during the period of experiment, maximum average radiation is 764 W/m^2 on 2nd April 2019 and minimum average solar radiation is 253.8 W/m^2 on 13th March 2019.

4.2 Open circuit voltage (V_{oc})

The deposition and accumulation of airborne dust significantly reduce the performance of the solar PV panels. As the dust deposited on the surface of PV panel, it covers solar cells and it doesn't allow the sunlight to pass through glass surface to solar cells. The dust covered solar cell doesn't participate to produce voltage; hence the total voltage production by solar panels decreases.

The variation in daily average V_{oc} of system A, system B and system C is shown in Fig 4.2(a), 4.2(b) and 4.2(c) respectively.

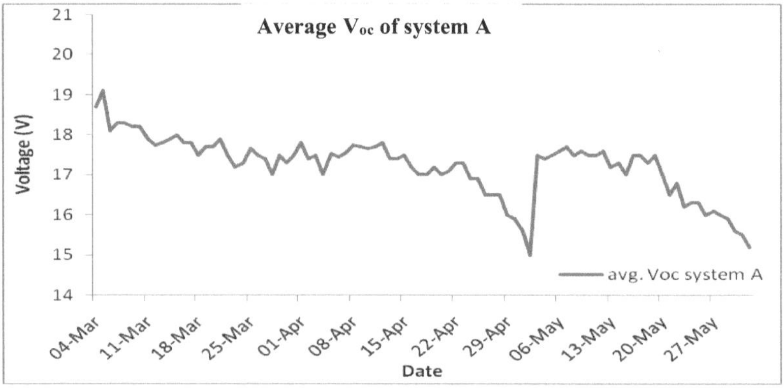

Figure 4.2 (a): Variation in daily average open circuit voltage for system A

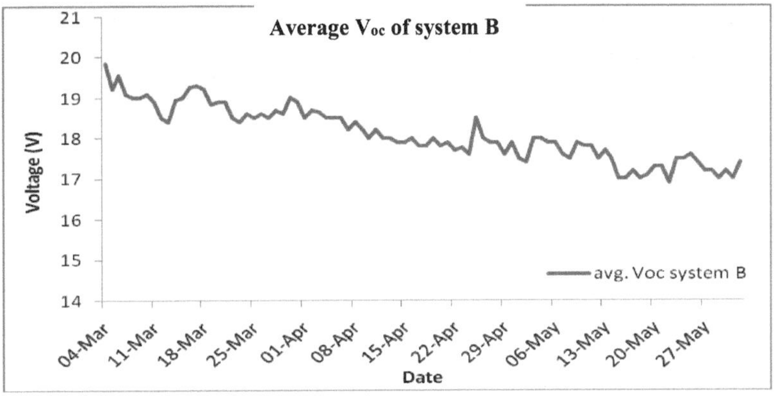

Fig 4.2 (b): Variation in daily average open circuit voltage for system B

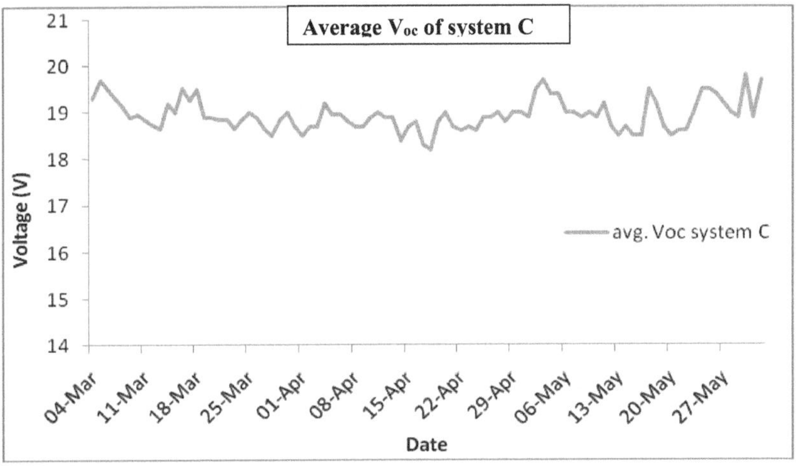

Fig 4.2 (c): Variation in daily average open circuit voltage for system C

It is observed from Fig. 4.2 (a) and Fig 4.2 (b) that as the time passes there is decrement in the open circuit voltage of system A and system B. The decrement is more in system A due to continuous deposition of dust on panel. There is a spike in voltage of system A at 3rd May because dust is removed by natural rain occurred. In system B, the panel is washed with water without any pressure. The water removes the dust particles to a large extent, but leaves smaller dust particles on the panel. There is a periodic decrement in voltage of system B due to periodic cleaning of system B. In Fig 4.2 (c) the variations is occurred due to periodic cleaning of panels by water mixed with surfactant. The voltage is consistent in system c because surfactant washing also removed small dust particles and allowed the full surface to expose to sun light.

4.3 Short circuit current (I_{sc})

The variation of the solar radiation affects significantly the short circuit current I_{sc}. Any change of the irradiation causes a proportional change in the short circuit current. The daily average short circuited current for system A, system B, system C is shown in Fig 4.3(a), Fig 4.3(b) and Fig 4.3 (c) respectively.

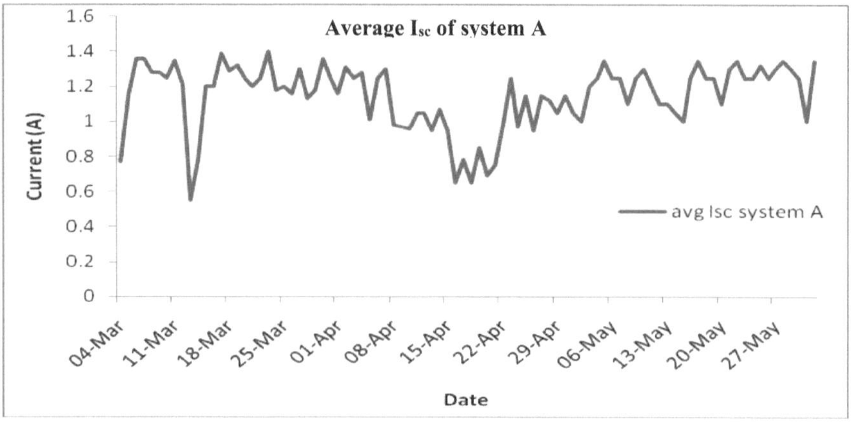

Figure 4.3 (a): Variation in daily average short circuited current for system A

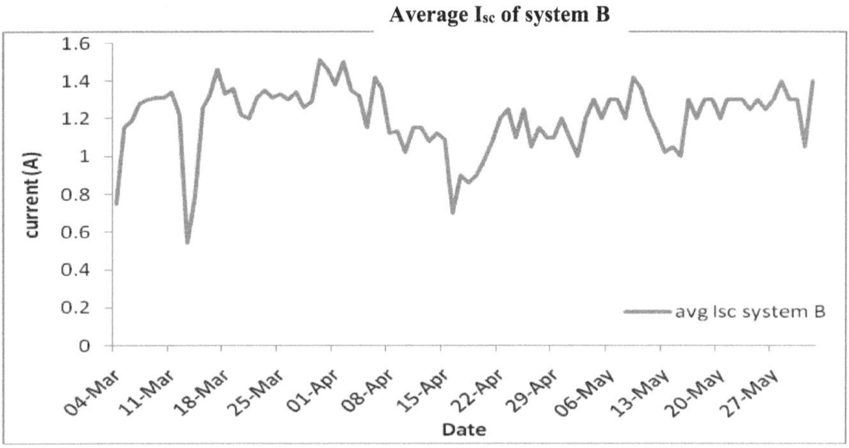

Fig 4.3 (b): Variation in daily average short circuited current for system B

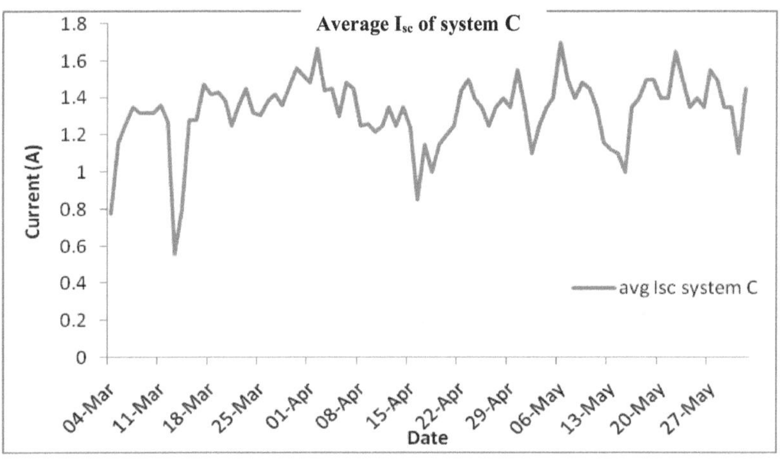

Fig 4.3 (c): Variation in daily average short circuited current for system C

It is observed that short circuit current varies with solar radiation i.e. as the radiation increases, the current increases in all three systems and if radiation decreases, current also decreases. The minimum current was measured at 3rd March due to minimum solar radiations.

4.4 Daily Maximum power generation (P)

Figure 4.4 represents the variation of daily maximum power generated by System A, System B and System C.

It is observed that daily average power generation of system A and system B is lower than power generation of system C. The power generation of system A is lower than that of system B because dust deposition in system A is more than that of system B. System C is generating consistent power because cleaning with surfactants doesn't allowed the dust to stick to its surface. Power generation in system A is increased on 3rd May because of natural rain removed dust from surface of system A.

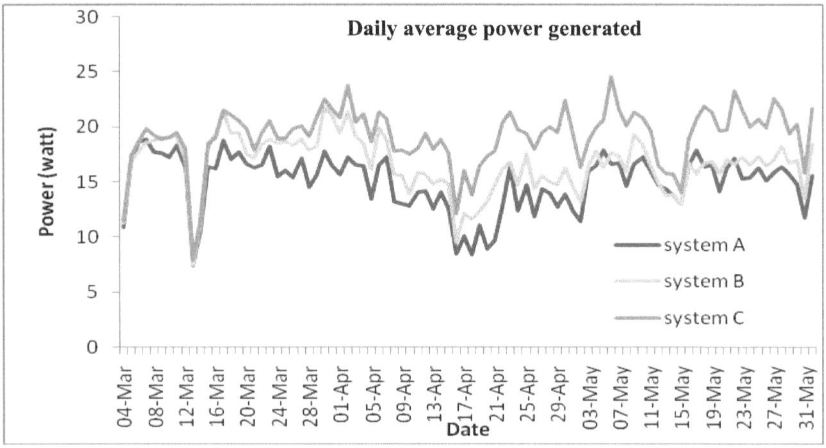

Figure 4.4 Daily average power generated by all three systems

4.5 Total maximum power generation

The total maximum power generation (in kWh) by all three systems during the experiment is shown in Fig 4.5. From figure; it was observed that the power generated by system A is 10.81 kWh in 90 days (March 04 to June 01, 2019). The system B generated 11.95 kWh and system C generated 13.78 kWh. This means the system C generated 27.5% more power than system A and 15.3% more power than system B in 90 days.

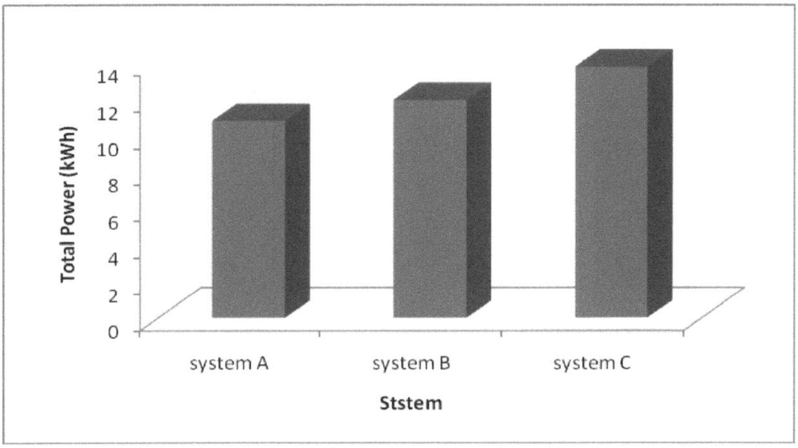

Figure 4.5 Total power generated during the experiment

4.6 Efficiency

The average daily efficiency of the system A, B and C is shown in Figure 4.6. From figure, it was observed that system C maintained its efficiency consistently while system A and system B, efficiency decreases due to dust deposition. The overall efficiency of System A was decreased by 29% and the efficiency of system B and system C was decreased by 18.17% and 5.2% respectively.

The efficiency of system B is decreased with a slope of 0.03/day. As the system A is not washed the efficiency of system A is decreasing significantly with a slope of 0.07/day.

Natural rain occurred at 3rd May a spike in efficiency of system A is clearly observed because system A is not washed manually. Rain water cleaned the surface of system A only up to certain extent.

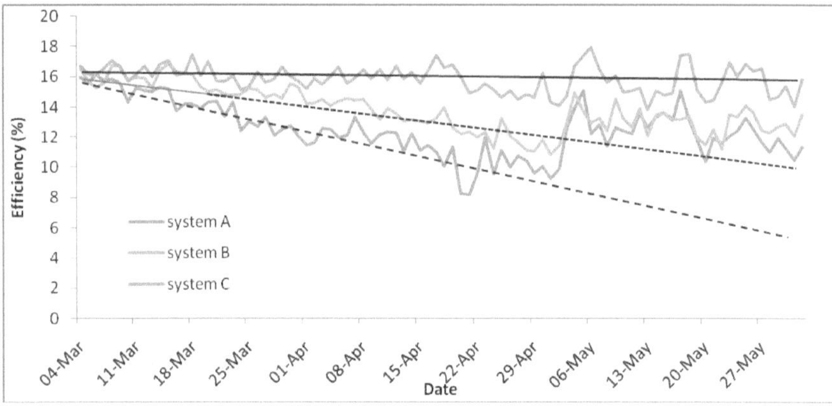

Figure 4.6 Total power generated during the experiment

4.7 Dust deposition on surface of panels

The airborne dust is deposited continuously on the PV panels. As the dust deposited, it reduces the performances of PV panels. Figure 4.7(a), 4.7(b) and 4.7(c) shows the surface of panels after 7 days, 30 days and 60 days respectively.

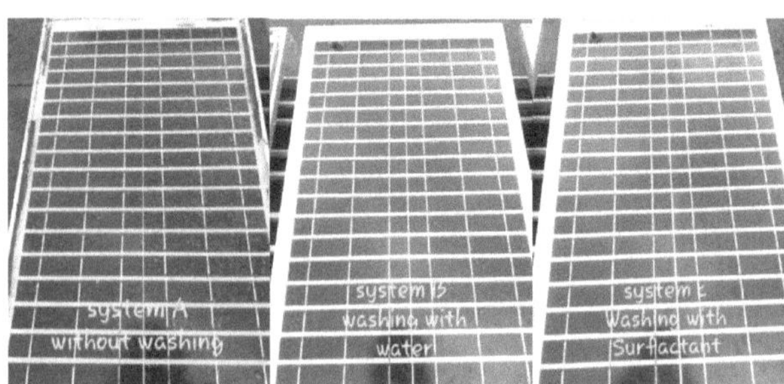

Figure 4.7 (a): Dust deposition after 7 days

After 7 days of operation a little dust is deposited on panel A while a little to no dust is deposited on panel B and on panel C there was no dust deposition.

Fig 4.7(b) Dust deposition after 30 days

After 30 days we found a thick dust layer on panel A while on panel B a very thin layer of dust is observed and panel C was clean and no dust particles were observed on surface of panel C.

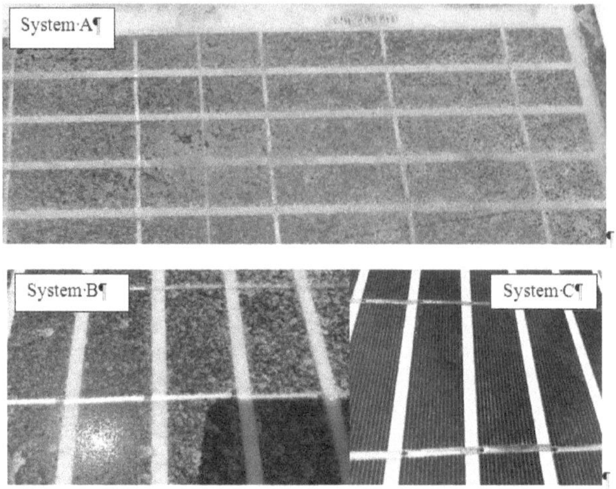

Fig 4.6 (c) Dust deposition after 60 days

After 60 days of operation a very thick layer of dust was observed on panel C while on Panel B a significant dust layer of very small dust particles with stains of water were observed and panel C was still clean.

4.8 Economic Survey

The cost for surfactants used for cleaning of system C can be found as follows:

Cost of surfactant / kg (INR) = 98

Total surfactant used in single wash = 1 gm

Total wash count in 90 days of experiment = 13

Total cost of surfactants during experiments = 98*0.001*13

= 1.274/-

Thus by spending INR 1.274, we are able to generate 27.5% more power in 90 days.

Chapter 5

CONCLUSIONS AND FUTURE SCOPE

This chapter deals with the conclusions drawn from the results and discussion and future scope.

5.1 Conclusions

The following conclusions are drawn from the experimental work:

The system C which was washed with surfactant maintained a consistent voltage because no dust deposition was found on system C. Small voltage drop found on system B on which a thin layer of dust is deposited and system A showed a significant drop in voltage as a thick layer of dust is found on surface of system A.

All systems i.e system A system B and system C showed the current variation directly proportional to the global solar radiation.

The total maximum power generation by system C washed by surfactant mixed with water was 13.78 kWh, and by system B and system A, it was 11.95 kWh and 10.81 kWh respectively. This means the system C generated 27.5% more power than system A and 15.3% more power than system B in 90 days.

The overall efficiency of system C washed with surfactant was higher than that of system B washed with water. The overall efficiency of System A was decreased by 29% and the efficiency of system B and system C was decreased by 18.17% and 5.2% respectively. Surfactant washing gives the consistent efficiency.

Dust deposition was found minimum on system C as the surfactant mixed with water removes all dust from panel. On system B, a thin dust layers was found on the surface of the panel and it was very sticky to the surface. Some spots of water were also found on the panel results restricted in incoming solar radiation on panel surface. A thick layer on the system A was also found as the panel was not washed.

The uses of surfactants seems feasible as the consumption of surfactants was very low as compared to increase in maximum power generation. By spending INR 1.274 in surfactants, we are able to generate 27.5% more power in 90 days.

5.2 Future scopes

There is a very extensive scope for future researchers to explore the use of different surfactants to enhance the performance of PV panels. The waste water after the cleaning of panels can be filtered and reused for panel cleaning. Other than cleaning with surfactants, self-cleaning coatings may be used for enhancing the performance.

Chapter 6

References

[1] K.A. Moharram, "Influence of cleaning using water and surfactants on the performance," Elsevier, 2013.

[2] (2019, july) ibef. [Online].https://www.ibef.org/industry/renewable-energy-presentation

[3] (2019, july) mnre india. [Online]. https://mnre.gov.in/physical-progress-achievements

[4] (2019, july) greenmatch. [Online].https://www.greenmatch.co.uk/blog/2015/09/types-of-solar-panels

[5] Patil P.A., Bagi J.S., Wagh M.M., Patil G.S., "design optimization, automation and testing analysis of dust cleaning mechanism for solar photovoltaic power plant," 2019.

[6] Thomas S.K., Joseph S., Sarrop T.S., Haris S.B., Roopak R. "Solar panel automated cleaning (SPAC) system,2018.

[7] Parrott B., Carrasco Zanini P., Shehri A., Kotsovos K., Gereige I. "Automated, robotic dry-cleaning of solar panels in Thuwal, Saudi Arabia using a silicone rubber brush," Elsevier, 2018.

[8] Kawamoto H., "Electrostatic cleaning equipment for dust removal from soiled solar panels," Elsevier, 2019.

[9] Deb D., Brahmbhatt N.L., " Review of yield increase of solar panels through soiling prevention, and a proposed water-free automated cleaning solution," Elsevier, 2018.

[10] Kawamoto H., Guo B., "Improvement of an electrostatic cleaning system for removal of dust from solar panels," Elsevier, 2018.

[11] Mazumder M.K., Horenstein M.N., Joglekar N.R., Sayyah A., Stark J.W., Bernard A.A.R., Garner S.M., Yellowhair J.E., Lin H.Y., Eriksen R.S., Griffin A.C., Gao Y., Centra R.L., Lloyd A.H., "Improvement of an electrostatic cleaning system for

removal of dust from solar panels," Elsevier, 2017.

[12] Sharma R., Wyatt C.A., Zhang J., Calle C.I., Mardesich N., Mazumder M.K.., Sarrop T.S., Haris S.B., Roopak R. " Experimental evaluation and analysis of electrodynamic screen as dust mitigation technology for future Mars missions,2009.

[13] Kawamoto H." Electrostatic transport and manipulation of lunar soil and dust,2008.

[14] Mazumder M.K., Sharma R., Biris A.S., Zhang J., Calle C., Zahn M.," Self-cleaning transparent dust shields for protecting solar panels and other devices,2007.

[15] Latthe S.S., Sutar R.S., Kodag V.S., Bhosale A.K., Kumar A.M., Kumar Sadasivuni K., Xing R., Liu S.," Self – cleaning superhydrophobic coatings: Potential industrial applications,2019.

[16] Mozumder M.S., Mourad A.-H.I., Pervez H., Surkatti R.," Recent developments in multifunctional coatings for solar panel applications: A review,2019.

[17] Zhan W., Wang W., Xiao Z., Yu X., Zhang Y," Water-free dedusting on antireflective glass with durable superhydrophobicity",2018.

[18] Adak D., Ghosh S., Chakraborty P., Srivatsa K.M.K., Mondal A., Saha H., Mukherjee R., Bhattacharyya R., "Non lithographic block copolymer directed self-assembled and plasma treated self-cleaning transparent coating for photovoltaic modules and other solar energy devices", 2018.

[19] Walz K.A., Shoemaker J.B., Scholes A.J., Jiang H., Silva J.M.S., Sanfilippo J., Zeltner W.A., Anderson M.A., " Experimental field trial of self-cleaning solar photovoltaic panels," 2018.

[20] Tao C., Zou X., Du K., Zhou G., Yan H., Yuan X., Zhang L., " Fabrication of robust, self-cleaning, broadband TiO_2–SiO_2 double-layer antireflective coatings with closed-pore structure through a surface sol-gel process," 2018.

[21] Arabatzis I., Todorova N., Fasaki I., Tsesmeli C., Peppas A., Li W.X., Zhao Z., " Photocatalytic, self-cleaning, antireflective coating for photovoltaic panels:

Characterization and monitoring in real conditions," 2018.

[22] Sutha S., Suresh S., Raj B., Ravi K.R., " Transparent alumina based superhydrophobic self–cleaning coatings for solar cell cover glass applications," 2017.

[23] Mehmood U., Al-Sulaiman F.A., Yilbas B.S., Salhi B., Ahmed S.H.A., Hossain M.K., "Superhydrophobic surfaces with antireflection properties for solar applications: A critical review,"2016.

[24] Ebert D., Bhushan B., "Transparent, superhydrophobic, and wear-resistant surfaces using deep reactive ion etching on PDMS substrate,"2016.

[25] Powell M.J., Quesada-Cabrera R., Taylor A., Teixeira D., Papakonstantinou I., Palgrave R.G., Sankar G., Parkin I.P., "Intelligent Multifunctional VO2/SiO2/TiO2 Coatings for Self-Cleaning, Energy-Saving Window Panels,"2016.

[26] Xu Q.F., Zhao Y., Kujan E., Liu J.C., Lyons A.M., "An anti-reflective and anti-soiling coating for photovoltaic panels," 2015.

[27] Ragesh P., Anand Ganesh V., Nair S.V., Nair A.S." A review on 'self-cleaning and multifunctional materials'",2014.

[28] He G., Zhou C., Li Z." Review of self-cleaning method for solar cell array," Elsevier, 2011.

[29] Verma L.K., Sakhuja M., Son J., Danner A.J., Yang H., Zeng H.C., Bhatia C.S., " Self-cleaning and antireflective packaging glass for solar modules," Elsevier, 2011.

[30] Williams R.B., Tanimoto R., Simonyan A., Fuerstenau S., " Vibration characterization of self-cleaning solar panels with piezoceramic actuation," Elsevier, 2007.

[31] Makkar A., Raheja A., Chawla R., Gupta S., " IoT Based Framework: Mathematical Modelling and Analysis of Dust Impact on Solar Panels," Elsevier, 2019.

[32] Yilbas B.S., Ali H., Yousaf M.R., Al-Sharafi A., "Hydrophobic Materials," Elsevier,

2018.

[33] Gaio J.N., de Campos K.M.A., Tiepolo G.M., " Determination of the ideal time for cleaning of photovoltaic panels aiming at better productivity - Case study of the implanted grid-connected photovoltaic system at the Federal University of Technology of Paraná," Elsevier, 2018.

[34] Nižetić S., Čoko D., Yadav A., Grubišić-Čabo F., " Water spray cooling technique applied on a photovoltaic panel: The performance response," Elsevier, 2016.

[35] Manisha mishra1, P.Muthuprasanna2*, K.Surya prabha2, P.Sobhita rani2, I . A.Satish babu3 , I. Sarath Chandiran4,G.Arunachalam5 and S.Shalini6. , " Basics and Potential Applications of Surfactants - A

Review," 2009.

[36] (2019, july) MNIT [Online].https://www.moniee.com/mnit

FSC
www.fsc.org
MIX
Papier | Fördert gute Waldnutzung
FSC® C083411

Kolibri 360 GmbH
Ferdinand-Jühlke-Straße 7
99095 Erfurt, Deutschland
produktsicherheit@kolibri360.de

Druck:
CPI Druckdienstleistungen GmbH
im Auftrag der
Kolibri 360 GmbH
Ferdinand-Jühlke-Str. 7
99095 Erfurt